Robotic Fish iSplash-II:
Realizing Fast Carangiform Swimming to Outperform a Real Fish

By
Richard James Clapham PhD

iSplash Robotics

Published by *iSplash* Robotics
www.isplash-robotics.com
r.j.c@ieee.org

Copyright © 2016
Published by *iSplash* Robotics 2016
Illustration by Richard James Clapham PhD

All rights reserved. No part of this publication may be reproduced, stored in a retrieval system or transmitted in any form or by any means, electronic, mechanical, photocopying or otherwise without the prior permission of *iSplash* Robotics.

ISBN-13: 978-1537270906
ISBN-10: 1537270907

Thank you for your purchase.

The Worlds Fastest Fish

Abstract—This paper introduces a new robotic fish, iSplash-II, capable of outperforming real carangiform fish in terms of average maximum velocity (measured in body lengths/second) and endurance, the duration that top speed is maintained. A new fabrication technique and mechanical drive system were developed, effectively transmitting large forces at high frequencies to obtain high-speed propulsion. The lateral and thrust forces were optimized around the center of mass, generating accurate kinematic displacements and greatly increasing the magnitude of added mass. The prototype, with a length of 32cm has significantly increased the linear swimming speed of robotic fish, achieving consistent untethered stabilized swimming speeds of 11.6BL/s (i.e. 3.7m/s), with a frequency of 20Hz.

Keywords: Robotic fish · Carangiform swimming · Mechanical drive system · Full-Body length.

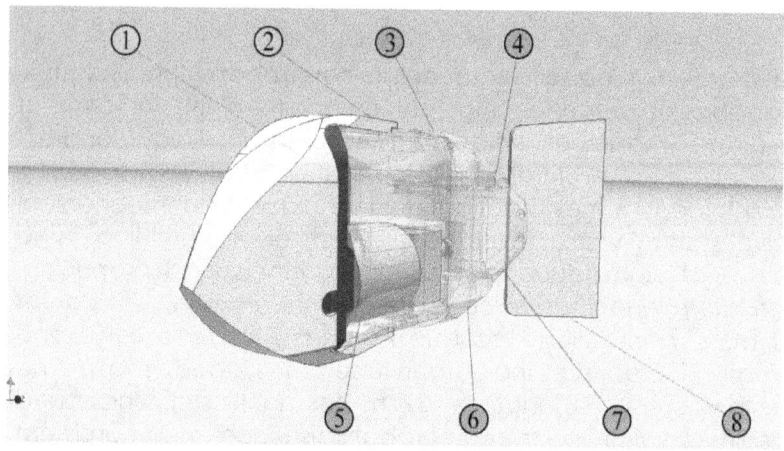

Figure 1. *iSplash*-II: 1-Anterior link; 2-Midbody transition links; 3- First posterior link; 4- Final posterior pivot; 5-Primary actuator; 6- Direct drive offset crank; 7- Tendon driven peduncle; 8-Compliant caudal fin.

I. INTRODUCTION

A. Background Description

To navigate through a marine environment, a robotic vehicle requires mobility to effectively contend with the physical forces exerted by the surrounding fluid. Live fish can coordinate their body motions in harmony with the surrounding fluid generating large transient forces efficiently, as opposed to rigid hull underwater vehicles (UV) powered by rotary propellers [1],[2],[3],[4]. For a man-made vehicle to achieve greater locomotive capability there is potential to engineer a structure that can accurately replicate the wave form of swimming fish.

Bainbridge's intensive observational studies measured live fish to attain an average maximum velocity of 10 body lenghts/ second (BL/s) [11]. A single high performance of a Cyprinus carpio was noted, achieving the swimming speed of 12.6BL/s (1.7m/s) with a stride rate of 0.7. Endurance at the highest velocities is limited, burst speeds can only be maintained for short durations of approximately one second. Velocities were measured to decrease to 7BL/s in 2.5s of swimming, to 5BL/s in 10s and to 4BL/s in 20s.

Although most work has focused on hydrodynamic mechanisms, current robotic fish are unable to gain the locomotive efficiencies of live fish, proving a complex challenge. There are two limitations in particular: (i) They cannot achieve accurate replication of the linear swimming motion as free swimming robotic fish generate kinematic parameter errors and therefore reduced propulsion; (ii) They have low force transfer due to the complexity of developing the powertrain, limited by mass, volume, force, frequencies and internal mechanical losses. Some examples of novel design approaches are Barrett's hyper-redundant Robotuna, achieving a maximum velocity of 0.65 body lengths/ second (BL/s) (0.7m/s) [5], Anderson's VCUUV with 0.5BL/s (1.2m/s) [6], Yu's discrete structure with 0.8BL/s (0.32m/s) [7], Essex's G9 with 1.02BL/s (0.5m/s) [8]; Wen's carangiform with 0.98BL/s (0.58m/s)

[9] and Valdivia y Alvarado's compliant method with 1.1BL/s (0.32m/s) [10]. The straight-line speed of current robotic fish, peaking at 1BL/s, is typically unpractical for marine based environments.

iSplash-I [12], a carangiform swimmer, (25cm, 0.35Kg), with an external power supply and formed of aluminum and steel, achieved a high-performance swimming motion. The developed novel mechanical drive system operated in two swimming patterns, a traditional posterior confined undulatory swimming pattern and the introduced coordinated full body length swimming pattern. The proposed swimming motion greatly improved the accuracy of the kinematic displacements and outperformed the posterior confined approach in terms of speed, achieving 3.4BL/s and consistently achieving a maximum velocity of 2.8BL/s at 6.6Hz with a low energy consumption of 7.68W.

It was noticed that throughout the field trials *iSplash*-I was able to replicate the key swimming properties of real fish. As frequencies were raised the prototype continued to increase velocity in both swimming modes. This matches Bainbridge's study of swimming fish, measuring no noticeable change in kinematics after tail oscillations are raised beyond 5Hz, indicating that only an altered frequency is required to increase swimming speed. Hence it is expected, that combining the critical aspects of the *iSplash*-I mechanical drive system with frequencies higher than 6.6Hz may significantly increase maximum velocity. In consideration of this, *iSplash*-II was developed, as shown in Fig. 1.

B. Research Objectives

The project aimed to achieve the fastest swimming speeds of live fish with seven main objectives: (i) to devise a prototype which operates in two swimming patterns, for further investigation of the carangiform swimming motion to be conducted; (ii) to significantly increase force transfer by achieving a high power density ratio in combination with an efficient mechanical energy transfer; (iii) to achieve unrestricted high force swimming by realizing a prototype capable of carrying a high powered energy supply; (iv) to develop a structurally robust mechanical drive system based on the critical properties proposed in [12], capable of intensively high frequencies of 20Hz; (v) to greatly reduce forward resistance by engineering a streamlined body considering individual parts' geometries and alignment throughout the kinematic cycle; (vi) to stabilize the free swimming prototype's unsteady oscillatory motion during intensively high frequencies to achieve a more efficient force transfer; (vii) to conduct a series of experiments measuring the prototype's achievements in terms of kinematic data, speed, thrust, and energy consumption in relation to driven frequency.

The remainder of the paper is organized as follows: Section **II** presents the investigated carangiform swimming patterns. Section **III** describes the new construction method. Section **IV** discusses the field trials undertaken and the experimental results obtained. Concluding remarks and future work are given in Section **V**.

II. Design Methodology

A. Mode 1: Traditional Approach

The kinematic swimming motion during linear locomotion of the Cyprinus carpio (common carp) is studied due to its high locomotive efficiency [13]. The selected carangiform applies the swimming method of body and/or caudal fin propulsion, identified by the portion of the body length actively displaced. The form of this propulsive segment, within the horizontal plane can be represented by a travelling wave. This body motion traditionally adopted in previous builds applies a rigid mid-body and anterior, concentrating the undulatory motion to the posterior end of the lateral length. Typically limited to <1/2 of the body length, the posterior propagating wave smoothly increases in amplitude towards the tail, consisting of one positive phase and one negative phase [4]. Described as Mode 1 and illustrated in Fig. 2, the posterior confined kinematics of the carangiform is of the form [5]:

$$y_{body}(x,t) = (c_1 x + c_2 x^2)\sin(kx + \omega t) \qquad (1)$$

where y_{body} is the transverse displacement of the body; x is the displacement along the main axis beginning at the nose; $k = 2\pi/\lambda$ is the wave number; λ is the body wave length; $\omega = 2\pi f$ is the body wave frequency; c_1 is the linear wave amplitude envelope and c_2 is the quadratic wave. The parameters $P = \{c_1, c_2, k, \omega\}$ can be adjusted to achieve the desired posterior swimming pattern for an engineering reference.

As previously mentioned, accurately matching the kinematic data of real fish is complex and free swimming robotic fish applying posterior confined displacements have shown kinematic parameter errors [8][10]. In particular, the lateral (F_L) and thrust (F_T) forces are not optimized. As a result, large anterior destabilization in the yaw plane is generated due to the concentration of posterior thrust, recoiling around the center of mass. Consequently the inaccurate anterior kinematics create significant posterior midline displacement errors. Hence, the linear locomotive swimming motion over the full length of body has large matching errors in comparison to real fish leading to reduced propulsive force and a higher cost of transport.

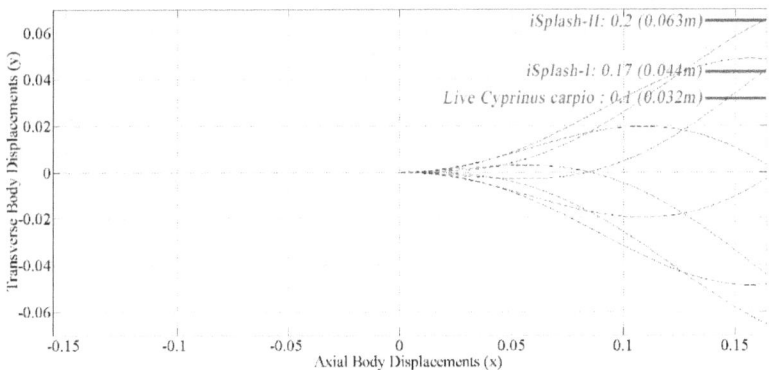

Figure 2. Mode 1: The wave form is confined to the posterior 1/2. (Parameters have been measured from experimental tests).

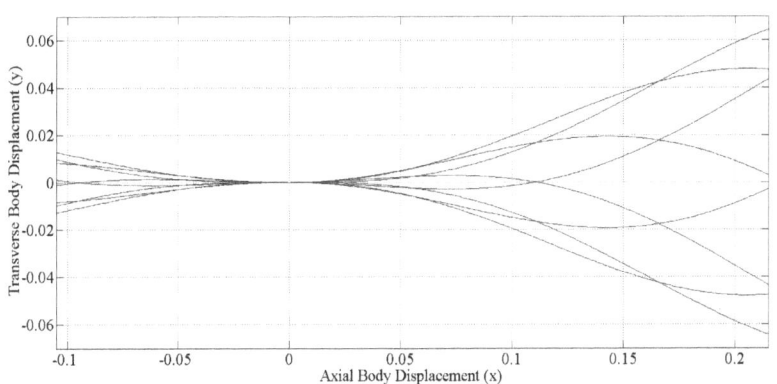

Figure 3. Mode 2: Showing the desired coordinated full-body kinematic parameters, generated based on the swimming motion of *iSplash*-I and the structural dimensions of *iSplash*-II.

B. Mode 2: Full-Body Swimming Pattern

Mode 2, illustrated in Fig. 3 is the full body carangiform swimming pattern of the *iSplash* platforms described in [12], which coordinates the anterior, mid-body and posterior body motions. This was based on intensive observation [12] and fluid flow theory [14],[15] which lead to a greater understanding of the carangiform swimming motion.

The Mode 2 displacements drive the anterior into the direction of recoil, reducing amplitude errors by optimizing the reaction force (F_R) of the propulsive elements. This enhances performance, increasing the magnitude of added mass by initiating the starting moment upstream, generating optimized F_L and F_T forces around the center of mass, increasing the overall magnitude of thrust contributing to increased forward velocity. A full description of the method of added mass can be found in [16]. Furthermore the developed body motion increases performance by allowing a smooth transition of flow along the length of the body, effectively coordinating and propagating the anterior formed fluid flow interaction downstream. We can extend the form of [8], an adaptation of (1) to generate the midline kinematic parameters of the full body displacements:

$$y_{body}(x,t) = \left(c_1 x + c_2 x^2\right)\sin\left(kx + \omega t\right) - c_1 x \sin\left(\omega t\right) \quad (2)$$

By evaluating the x location at the center of mass, measured optimal at 0.15-0.25 of the body length and tail amplitude at 0.1 [13], the relationships between the defined parameters P ={0.63,0,21.6,8} shown in Fig. 3 can be found.

The full-body swimming motion of *iSplash*-I reduced kinematic matching errors over the full body length. Mode 2 was found to reduce the head amplitude by over half, from 0.17 (0.044m) in Mode 1 to 0.07 (0.018m). The tail amplitude was measured to increase performance with larger values than the common carp at 0.1. Both Modes were able to attain values of 0.17 (0.044m) due to achieving anterior stabilization.

The location of the center of mass was improved and close to the optimum range, Mode 2 with a reduced error location of 0.33 in comparison to Mode 1 of 0.5. Reviewing the kinematic data we can see, that the prototype achieved high kinematic accuracy producing a low cost of transport. In consideration of this, we aimed to precisely replicate its swimming motion parameters.

III. New Construction Method

A. Mechanical Design

In order to increase the swimming speed a new mechanical drive system was required, able to effectively transmit large forces at high tail oscillation frequencies. In consideration of this, a feasible design structure to fit the linear swimming patterns of both modes was developed. A powertrain deploying a single motor with continuous rotation was developed, more complex to devise without internal mechanical loss [5], it is advantageous in comparison to multilink servos or smart materials, which are limited by force, frequency, volume and mass distribution [7],[8],[9].

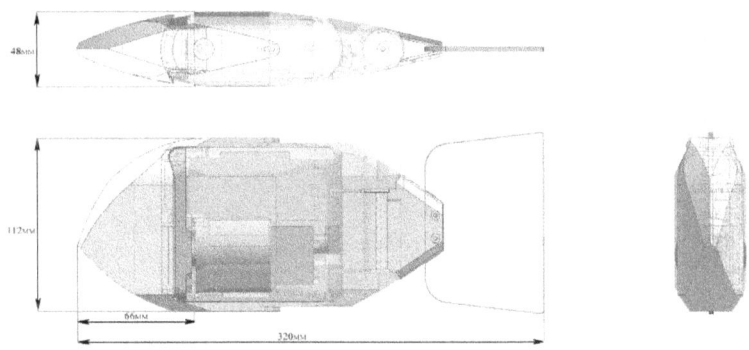

Figure 4. 1-Plan; 2-Side 3-Front view.

As the build required a high power density ratio, the structural arrangement was governed by the dimensions of the large electrical motor 83mm long x 50mm diameter. This required a slight increase in body length from 250mm to 320mm and a significant adjustment to the link structure to take the mass of the actuator into consideration, removing the coupled mid-body joint and the associated discrete linkages. The discrete construction method, defined as a series of links or N links, aims to achieve accurate midline kinematic parameters whilst minimizing complexity of the mechanical drive and linkages. The sequence of links can generate the required swimming motion by locating the joints to the spatial and time dependent body wave. The fully discretized body wave fitting method is given in [7], [8].

The assembly of *iSplash*-II is illustrated in Fig. 6, showing the four joints distributed along the axial length. Three rigid links are coupled to a compliant fourth link and caudal fin with stiffness distribution, devised to generate a smooth body to tail transition phase of the posterior undulations. The developed modular build allowed for both Modes of operation to be applied to the same prototype by adjusting the configuration. Links III and IV are actuated to generate the posterior kinematics of operational Mode 1, Mode 2 actuates all links along the axial length to provide anterior, mid and posterior body displacements.

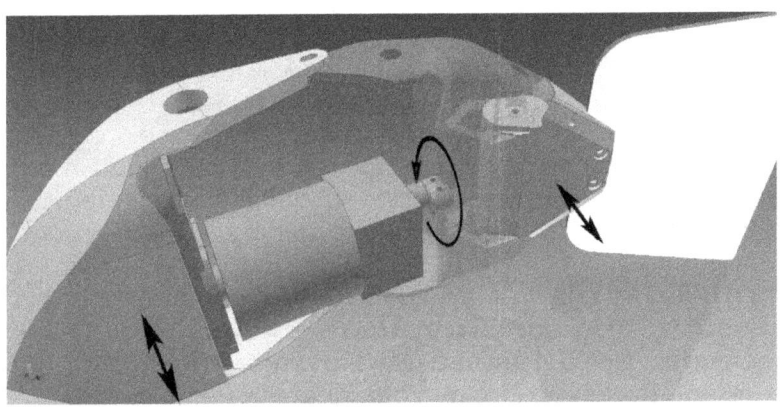

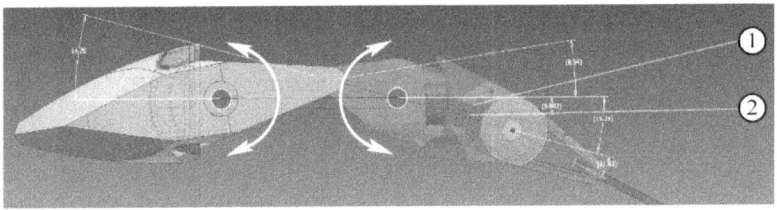

Figure 5. Power transmission system: 1-Offset crank; 2-Posterior link.

It was proposed in [12], that the outer profile of the coordinated full-body swimming pattern, represented by the aerofoil section NACA (12)520 aids the fluid flow interaction, producing greater locomotive speeds. In consideration of the simplified link assembly and estimated center of mass, the head and tail amplitudes were increased. We can see in Fig. 6 that the approximation of a traveling wave using link end points I-IV and turning angles of joints 1-3 of the reduced link arrangement provides an accurate curve alignment agreeable with the form of (2), therefore reducing errors and excrescences in the outer profile and achieving accuracy with the required aerofoil section.

B.   Power Transmission System

The leading tail discrete link III is directly driven by the single bearing crank shaft attached to the output shaft of the primary actuator, increasing power distribution to the posterior. As link III is actuated, link IV is passively displaced. This final posterior linkage IV, coupled to the compliant caudal fin is anchored by 4 expandable tendons attached to the main chassis rear bulkhead, crossing through linkage III. The anterior link I is transmitted motion by paired linkages fixed at points P_5 and P_6, located at the top and bottom of the main chassis. The developed mechanical design required precision fitment of the chassis, crankshaft, cantilevers and linkages to reduce internal mechanical losses, avoid deadlock and reduce friction.

Illustrated in Figs. 5, 6 and 7 is the developed powertrain transmitting rotary power to linear oscillating links. All driven link amplitudes are determined by the single offset crank. L_3 represents the leading tail discrete link of the structure.

The maximum amplitude of the link length L_3 and L_1 at point P_2 and P_4 respectively are determined by the predetermined maximum crank offset P_1. The coordinates of P_1 (P_{1x}, P_{1y}), P_2 (P_{2x}, P_{2y}), P_3 (P_{3x}, P_{3y}) and P_4 (P_{4x}, P_{4y}) can be derived by:

$$\begin{cases} P_{1x} = A+B \\ P_{1y} = (A+B)\tan\theta_1 \end{cases} \quad \begin{cases} P_{2x} = P_{1x} + C\cos\theta_1 \\ P_{2y} = P_{1y} + C\sin\theta_1 \end{cases} \quad (3)$$

$$\begin{cases} P_{3x} = -F\cos\theta_1 \\ P_{3y} = -F\sin\theta_1 \end{cases} \quad \begin{cases} P_{4x} = -(L_2 + D) \\ P_{4y} = -P_{3y}D/E \end{cases} \quad (4)$$

The length of L_1 can be derived by $L_1^2 = P_{4x}^2 + P_{4y}^2$. Assume that ω_1 is the angular velocity of the link L_1, and the velocity vector V_{P4} is perpendicular to L_1. We have:

$$\begin{cases} V_{p4x} = -\omega L_1 \sin\theta_3 \\ V_{p4y} = \omega L_1 \cos\theta_3 \end{cases} \quad (5)$$

where V_{p4x} and V_{p4y} are the decomposed vectors of the velocity vector $V_{P4} = \omega_1 L_1$.

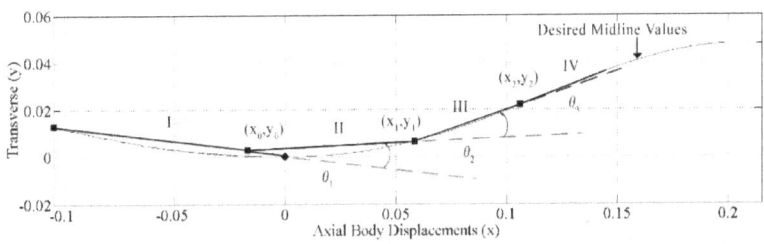

Figure 6. Link approximation, illustrating accurate kinematic matching;

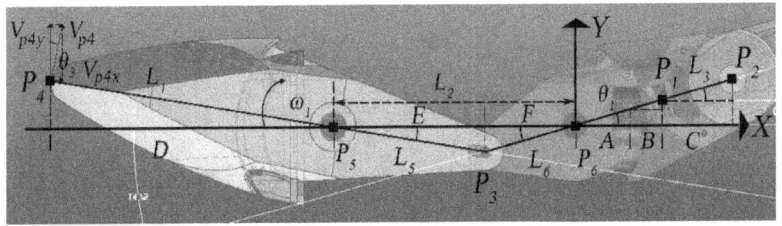

Figure 7. Schematic drawing of the offset drive crank and linkages;

C. C. Fabrication

The prototype *iSplash*-II is shown in Fig. 8 with the physical specifications given in Table I. The entire body was digital modeled and formed using 3D printing techniques, at layers of 0.09mm in PLA filament. This method produced precise 3D structural geometries of the individual segments and pre-determining alignment tolerances throughout the complete kinematic cycle. It was a key challenge to develop a high power density build, small in size with high structural strength. The individual printed parts were optimized for robustness through physical strength tests and computational stress analysis, highlighting initial areas of weakness. These parts were re-printed many times in order to realize high frequency actuation. As PLA filament has a low melting point softening at approximately 60°C, material wear at the pivots and actuated surfaces was reduced by acetal bushes and inserts, at the cost of additional weight.

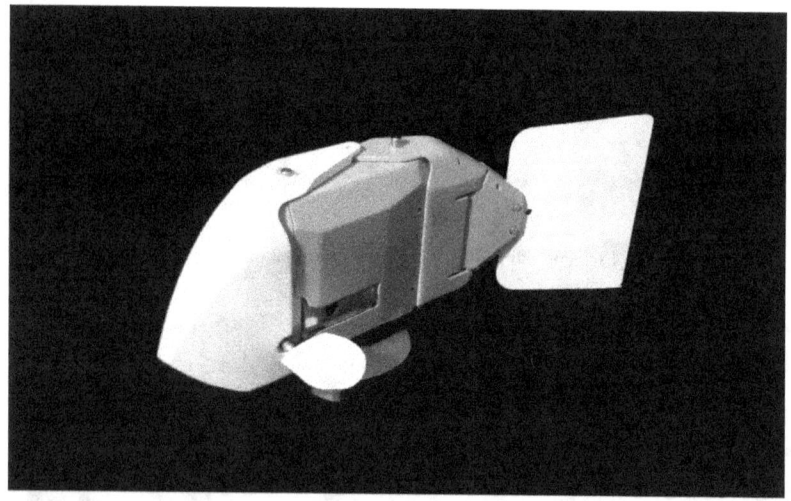

Figure 8. *iSplash*-II

It was necessary for the body size to be compact, as increasing the build geometric magnitude will increase the resistance during forward motion and therefore the power consumption required [5],[12]. An accurate approximation of the streamlined body shape of the common carp was achieved within the horizontal plane illustrated in Fig. 4. The maximum thickness of the cross section is measured optimal at 0.2 of the body length [13] and was favorably positioned therefore reducing pressure drag.

The static stability in the horizontal and vertical planes is affected by material density distribution. For linear locomotive research open loop stability is beneficial, this was achieved by the relative position of buoyancy being higher than the center of mass, as the surrounding fluid counterbalances the gravitational weight [17]. The short body length greatly increased the difficulty in achieving open loop stability as the finest weight change in structure of individual pieces distributed across the assembly dramatically affected stability and buoyancy. This was solved by collaborating the individual parts of the modular build by adjusting the geometries and the inner structure's weight to strength configuration.

The prototype was designed with increased stability in roll and pitch as the large mass of the electric motor, 0.6kg, 75% of the total mass, was positioned low within the structure. To achieve a short body length, contain the embedded system and 11.1V LiPo power supply and counteract the large mass of the primary actuator the build volume was increased vertically. This aided stability, as the lightweight PLA material and increased height positioned the center of buoyancy at the top of the prototype.

Mobility within the vertical plane was achieved to maintain a stable mid tank trajectory during free swimming. Two rigid morphological approximations of pectoral fins were developed and positioned at the leading bulk head of the main chassis, actuated by a single servo motor. A cross beam anchored on both sides of the centralized motor was formed to link, support and actuate the control surfaces. The addition of pectoral fins required a compact mechanism to be devised due to the very restrictive space available.

PHYSICAL PARAMETERS OF *iSPLASH*-II

Parameter	Specific Value
Body Size: m (LxWxH)	0.32 x 0.048 x 0.112
Body Mass: Kg	0.835
Actuator:	Single electric motor
Actuator Mass: Kg	0.63
Power Supply:	11.1V onboard LiPo battery
Manufacturing Technique:	3D Printing
Materials:	PLA Filament, Acetal, Stainless
Primary Swimming Mode:	Linear Locomotion
Additional Maneuverability:	Vertical plane
Additional Control Surfaces:	Pectoral fins
Caudal Fin Material:	Polypropylene
Thickness of Caudal Fin: mm	2.3
Caudal Fin Aspect Ratio: AR	1.6

IV. Experimental Procedure and Results

A. Field Trials

A series of experiments were conducted in order to verify the prototype by evaluating the locomotive performance of Modes 1 and 2 in terms of kinematic parameters, speed, force and energy consumption at frequencies within the range of 5-20Hz. It was required that the measurements were averaged over many cycles to increase the accuracy of data, once consistency of operation was achieved and stabilized free swimming was obtained. The test results are summarized in Table II. Experiments were conducted within a test tank, 5m long x 2m wide x 1.5m deep. Free swimming between two fixed points at a distance of 4m was used to evaluate maximum speed. The prototype had sufficient space to move without disturbances from side boundaries and the free surface, capable of consistent untethered swimming at mid height of the tank aided by adjusting the angle of pectoral fins during swimming.

Locomotion at high speeds was unachievable without extensive stability optimization. Once achieved, an accurate straight line trajectory was possible. A thorough description of the improvements undertaken on the mechanical structure and the extent of the intensive destabilization are beyond the scope of this paper. In addition, the devised mechanical drive system was found to be very robust, showing no signs of structural failure throughout the field trials whilst actuating at high frequencies over long periods and accidentally hitting the walls of the test tank.

B. Swimming Pattern Observation

The frame sequence of Mode 2 in eight instances, at time intervals of 0.006s throughout one complete body cycle at

19Hz is illustrated in Fig. 9. The obtained midline was tracked at 50 frames per second and is plotted against the desired amplitude envelopes of the anterior and posterior from Fig. 2 for comparison.

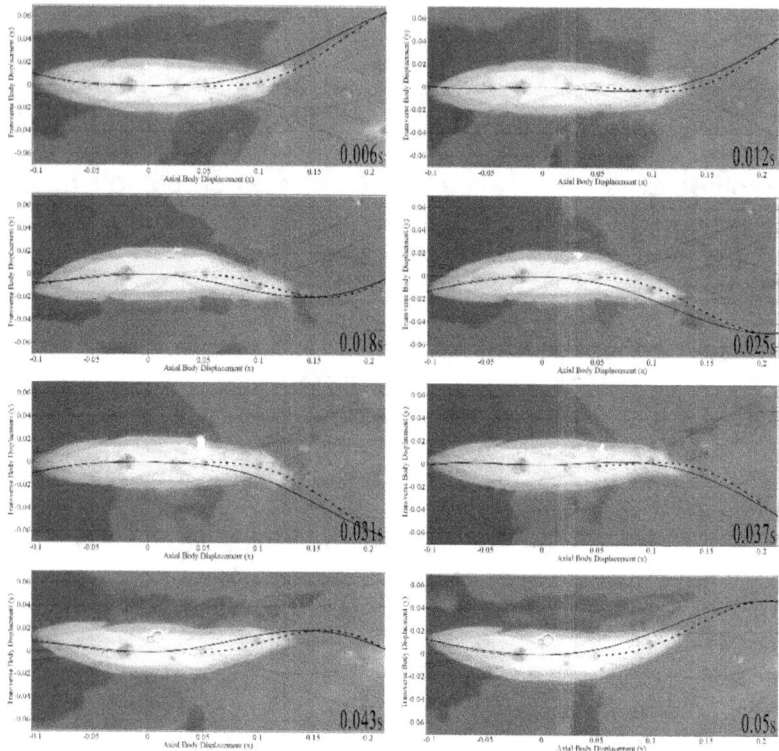

Figure 9.   Frame sequence of Mode 2 during one full body cycle, eight instances every 0.006s. The desired midline (—) (illustrated in figure 3) and the generated kinematics (illustrated in figure 2) from locomotion at 20Hz (- -) are shown for comparison. We can see the prototype achieved an anterior amplitude value of 0.04L (0.013m) and a tail amplitude value of 0.20L (0.063m).

When observing the midline of Mode 2 it can be seen, that the desired full body coordination presented in [12], was not achieved. As previously described in Section III-A the build required a simplified link structure due to power density constraints. Although the estimated midline curve alignment tested during stationary actuation was accurate, the excessive mass of the primary actuator held the main chassis (the entire length of link II) fixed in line with the forward heading and no single pivot point was obtained. Consequently, the swimming motion during locomotion was found to produce matching errors over the full-body in comparison to the desired swimming pattern of *iSplash*-I.

Comparing both Modes taking into consideration that the mid-body was held rigid, the anterior amplitude of Mode 2 was measured to be 0.04 (0.013m) of the body length, equivalent to the common carp, whereas Mode 1 was found to generate <0.01 (0.003m) head amplitude. In addition, the large centralized mass arrangement and increased depth of body effectively minimized recoil forces and aided the stability of the posterior, allowing for accurate posterior amplitude and large thrust forces to be generated.

It can be seen, that the developed posterior structure can accurately mimic the undulatory parameters of real fish, as the components of link IV can be adjusted experimentally to provide the targeted midline during free swimming at various frequencies. Both Modes were able to generate accurate amplitudes of 0.1 of the body length and attain large tail amplitudes of 0.2 (0.063m) which was found to significantly increase performance. This value is twice the size of the observed value of the common carp at 0.1 and is increased over the first generation at 0.17. This generated amplitude is greater than the highly efficient swimming motion of a dolphin measured at 0.175 [13].

The caudal fin was formed with a low aspect ratio (AR). Although not yet thoroughly investigated, this tail was measured to achieve the highest maximum velocity and acceleration during the initial field trails. AR is defined as: $AR=b^2/Sc$ where b squared is the fin span and Sc is the projected fin area. In this case the AR was 1.6.

COMPARISON OF TEST RESULTS BETWEEN MODES 1 & 2

Parameters	Mode 1	Mode 2
Maximum Velocity: BL/s (m/s)	11.6 (3.7)	11.6 (3.7)
Acceleration time to Max Velocity: s	0.6s	0.6s
Frequency: Hz	20	20
Reynolds Number: Re (10^6)	1.2	1.2
Strouhal Number: St	0.34	0.34
Maximum Thrust: N	9	9
Max Power Consumption Air: W	120	120
Max Power Consumption Water: W	120	120
Swimming Number: Sw	0.58	0.58
Head Swing Amplitude: m	0.003	0.013
Tail Swing Amplitude: m	0.063	0.063
Body length displaced: %	51	76

C. Experimental Results

In Fig. 10 the average energy economy in relationship to driven frequency is shown, comparing both operational Modes in air and water. It can be seen that both Modes actuating in water consumed a maximum 120W at 20Hz. This measurement was obtained by a connecting tethered power supply and no noticeable increase in energy consumption was measured due to a resistance of the surrounding liquid. The result of high energy consumption can be greatly improved as the tests indicated large mechanical gains when actuating the mechanical drive system without link IV, improving from a 120W to 70W consumption. This was a result of pressure increase at higher velocities, as link IV was actuated, the tendons were required to be tighter to provide the desired posterior kinematics, putting increased strain on the mechanism. Despite the high energy consumption, the prototype can maintain an operational time of approximately ten minutes at maximum velocity (estimated by video recording multiple runs), far surpassing the endurance of live fish, as equivalent burst speeds can only be maintained for short times of around one second. We can assume that engineering a greater mechanically efficient drive of link IV in the next generation may significantly improve endurance, relating to an estimated reduced energy consumption of ~50%.

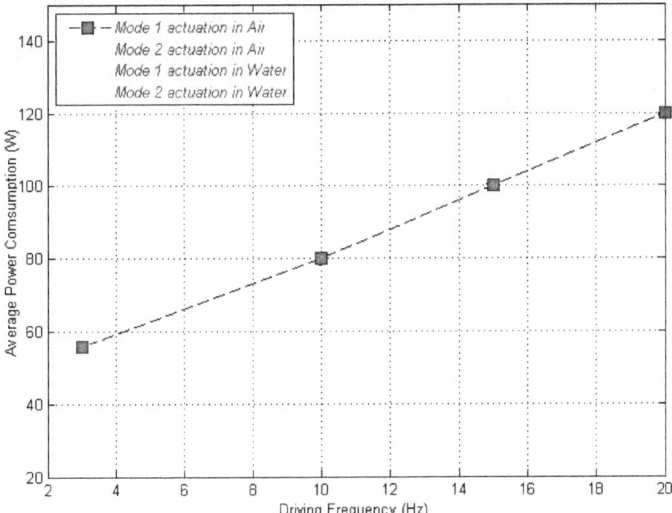

Figure 10. Comparison of average electrical power consumption over driven frequency of both Modes, actuating in air and water. No noticeable change was measured during testing for both modes in air and in water.

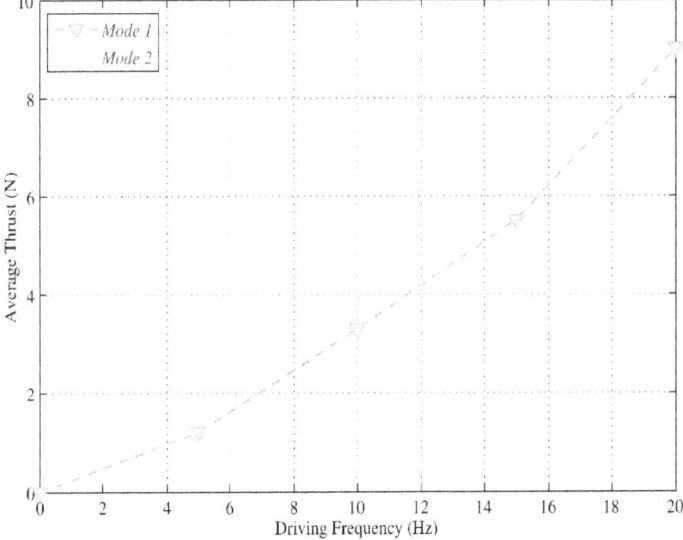

Figure 11. Comparison of average thrust in relationship to driven frequency. Modes 1 and 2 have equivalent measurements.

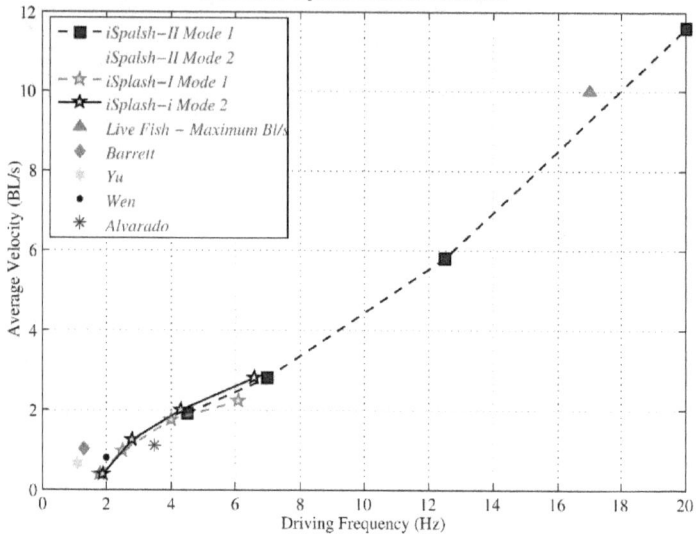

Figure 12. Comparison of average velocities achieved by both Modes, against robotic and live fish. Modes 1 and 2 measured equal velocities.

As illustrated in Fig. 11, the developed build with a high power density ratio can generate a great amount of force of up to 9N. This can be effectively transferred in the water, accelerating both Modes to maximum velocity in approximately 0.6s. The relationship between velocity (speed divided by body length) and driven frequency is shown in Fig. 12. The corresponding values of Modes 1 and 2 during consistent swimming are shown and compared to current robotic fish. Both operational Modes can achieve an average maximum velocity of 11.6Bl/s, (i.e. 3.7m/s) at 20Hz, increasing performance in comparison with *iSplash*-I and current published robotic fish which typically peak around 1BL/s. This result also outperforms the average maximum velocity of real fish measured at 10BL/s. The values illustrated in Fig. 12 show that applying the operational Mode 2 swimming pattern had no effect on performance due to kinematic alignment errors, discussed in Section IV-B, therefore it is predicted that the magnitude of added mass in both modes is equal. Hence, we can estimate that accurately applying the coordinated full-body swimming pattern of *iSplash*-I may increase speed by a further 27%.

A prominent parameter for analyzing BCF locomotive performance is the Strouhal number (St), defined as $St=fA/U$, where f denotes the frequency, A denotes the tail amplitude and U is the average forward velocity. St is considered optimal within the range of $0.25 < St < 0.40$ [15]. The measured $St = 0.34$ under the condition of $Re = 1.2 \times 10^6$, in both Modes is within the desired range. The prototypes Swimming number (Sw) (distance travelled per tail beat) is highly efficient, measuring a Sw of 0.58 in comparison to the previous build with a Sw of 0.42 and close to the particular efficient common carp with a Sw of 0.70 [11],[13].

We have undertaken experiments to gain knowledge if raising driven frequencies greater than the previous build of 6.6Hz would continue to increase speed without peak or decline. This was achieved measuring a continued increase in velocity up to intensively high frequencies of 20Hz. Mimicking the swimming properties of real fish, frequency has become the key variable to enhance the linear locomotive performance of the *iSplash* platforms.

V. Conclusion and Future Work

This paper describes the development and experimental analysis of *iSplash*-II. The study aimed to realize the fastest speeds of live fish. A high-performance prototype was developed, robust, compact, naturally buoyant, carrying its own power supply, with a high power density and able to effectively transmit large forces at intensively high tail oscillation frequencies for untethered high-speed propulsion.

Although the desired kinematics over the full body could not be attained due to the power density requirements (with the primary actuator 75% of the total mass), the devised assembly was able to reduce the recoil around the center of mass, therefore generating an effective propulsive mechanism. As a result, large posterior forces and tail amplitudes 0.2 of the body length (with smooth generated undulations from mid-body to tail tip) were attained. The prototype was able to accelerate to steady state swimming in an approximate time of 0.6s, maintain an endurance at maximum speed for approximately ten minutes (greater than the measurement of real fish of approximately one second), realize a highly efficient stride rate (Sw) and attain high tail oscillatory frequencies without early peak, decline or mechanical failure.

iSplash-II, a 32cm untethered carangiform swimmer, 0.835kg, formed in PLA filament, consistently achieved a maximum velocity of 11.6BL/s (i.e. 3.7m/s) at 20Hz with a stride rate of 0.58 and a force production of 9N. Capable of outperforming the recorded average maximum velocity of real fish measured in BL/s, attaining speeds adequate for real world environments

Our future research will focus on the following aspects to further improve the swimming performance: (i) continue to raise driven frequency to achieve greater speeds over the fastest real fish. As the build showed no signs of failure an initial aim of 40Hz can be made; (ii) to accurately emulate the kinematic parameters of the full-body swimming motion [12], indicating that maximum velocity will increase a further 27%; (iii) to replace the drive mechanism of link IV, to significantly improve the energy consumption; (iv) to optimize the tail amplitude, shape, 3D deformation and magnitude; (v) to apply the behavioral technique of burst and coast, as live fish generating 10BL/s at the burst stage reduce the cost of transport by approximately 50% [18]; (vi) to develop mobility within the horizontal plane with estimated turning diameter of < 1L.

Acknowledgments

Our special thanks go to Richard Clapham senior for his technical assistance towards the project.

References

[1] P. R. Bandyopadhyay, "Maneuvering hydrodynamics of fish and small underwater vehicles," Integr. Comparative Biol., vol. 42, no. 1, pp. 102–17, 2002.

[2] G. S. Triantafyllou, M. S. Triantafyllou, and M. A. Grosenbaugh, "Optimal thrust development in oscillating foils with application to fish propulsion," J. Fluids Struct., vol. 7, pp. 205–224, 1993.

[3] J.J. Videler, "Fish Swimming", Chapman and Hall, London, 1993.

[4] J. Lighthill, "Mathematical Biofluiddynamics", Society for Industrial and Applied Mathematics, Philadelphia, 1975.

[5] D.S. Barrett, M.S. Triantafyllou, D.K.P. Yue, M.A. Grosenbaugh, and M. J. Wolfgang, "Drag reduction in fish-like locomotion," J. Fluid Mech., vol. 392, pp. 183–212, 1999.

[6] J.M. Anderson, and N.K. Chhabra, "Maneuvering and stability performance of a robotic tuna", Integrative and Comparative Biology, Vol: 42, iss: 5, pp: 1026-1031, Nov 2002.

[7] J. Yu, M. Tan, S. Wang and E. Chen. "Development of a biomimetic robotic fish and its control algorithm," IEEE Trans. Syst., Man Cybern. B, Cybern,34(4): 1798-1810, 2004.

[8] J. Liu and H. Hu, "Biological Inspiration: From Carangiform fish to multi-Joint robotic fish," Journal of Bionic Engineering, vol. 7, pp. 35–48, 2010.

[9] L. Wen, G.H. Wu, J.H Liang and J.L Li, "Hydrodynamic Experimental Investigation on Efficient Swimming of Robotic Fish Using Self-propelled Method", International Journal of Offshore and Polar Engineering, Vol.20, pp. 167~174, 2010.

[10] P. Valdivia y Alvarado, and K. Youcef-Toumi, "Modeling and design methodology for an efficient underwater propulsion system", Proc. IASTED International conference on Robotics and Applications, Salzburg 2003.

[11] R. Bainbridge, "The Speed Of Swimming Of Fish As Related To Size And To The Frequency And Amplitude Of The Tail Beat", J Exp Biol 35:109–133, 1957.

[12] R.J. Clapham and H. Hu, "*iSplash*-I: High Performance Swimming Motion of a Carangiform Robotic Fish with Full-Body Coordination," Accepted for 2014 IEEE International Conference on Robotics and Automation, May 31 - June 7, 2014, Hong Kong, China.

[13] M. Nagai. "Thinking Fluid Dynamics with Dolphins," Ohmsha, LTD, Japan, 1999.

[14] M. W. Rosen, "Water flow about a swimming fish," China Lake, CA, US Naval Ordnance Test Station TP 2298, p. 96, 1959.

[15] M.J. Wolfgang, J.M. Anderson, M.A. Grosenbaugh, D.K. Yue and M.S. Triantafyllou, "Near-body flow dynamics in swimming fish," September 1, 1999, J Exp Biol 202, 2303-2327

[16] P.W. Webb, "Form and function in fish swimming," *Sci. Amer.*, vol. 251, pp. 58–68, 1984.

[17] G.V. Lauder and E.G. Drucker, "Morphology and Experimental Hydrodynamics Of Fish Control Surfaces," IEEE J. Oceanic Eng., Vol. 29, Pp. 556–571, July 2004.

[18] J.J. Videler and D. Weihs, "Energetic advantages of burst-and-coast swimming of fish at high speeds," J. Exp. Biol., 97:169-178, 1982.

iSplash Robotics

Published by *iSplash* Robotics UK
www.isplash-robotics.com
Copyright © 2016

Published by *iSplash* Robotics 2016
Illustration by Richard James Clapham PhD

All rights reserved. No part of this publication may be reproduced, stored in a retrieval system or transmitted in any form or by any means, electronic, mechanical, photocopying or otherwise without the prior permission of Natural Classics.

ISBN:

Thank you for your purchase.

www.ingramcontent.com/pod-product-compliance
Lightning Source LLC
Chambersburg PA
CBHW071839200526
45169CB00020B/1988